GW00471344

Geology of the Leadhills district

a brief explanation of the geological map Sheet 15E Leadhills

J D Floyd

abridged from the Sheet Description *by* E A Pickett

Bibliographic reference

FLOYD, J D. 2003. Geology of the Leadhills district —
a brief explanation of the geological map.
Sheet Explanation of the British Geological Survey.
1:50 000 Sheet 15E Leadhills (Scotland).

Keyworth, Nottingham: British Geological Survey

CONTENTS

Notes

The word 'district' used in this account refers to the area of Sheet 15E Leadhills. National grid references are given in square brackets. Most of the Leadhills district lies in 100 km square NS, with a small strip along the southern margin falling in square NX. Therefore, unless otherwise noted, all grid references used in this description should be prefixed by NS. Symbols in round brackets after lithostratigraphical names are the same as those used on the 1:50 000 geological map. Numbers referred to in plate captions are BGS registration numbers.

Acknowledgements

This *Sheet Explanation* was compiled by E A Pickett, Regional Editor, Integrated Geoscience Surveys (Northern Britain), Edinburgh, and is based on the approved version of the *Sheet Description* for the Leadhills district authored by J D Floyd. Full acknowledgements are to be found within the *Sheet Description*.

The grid, where it is used on figures, is the National Grid taken from Ordnance Survey mapping.

© Crown copyright reserved Ordnance Survey licence number GD272191/2003.

1 Introduction

The Leadhills district lies at the northern margin of the Southern Uplands. It straddles the Southern Upland Fault and includes a small area of the Midland Valley in the north-west of the district. The Southern Uplands Terrane, to the south of the fault, is characterised by rolling hills of Lower Palaeozoic rocks, with several Carboniferous and Permian outliers. The Midland Valley Terrane consists mainly of Siluro-Devonian sandstone and minor lava flows, and includes the southern part of the formerly important Douglas Coalfield. The district is largely rural, with the main population centres in the villages of Abington and Crawford in the valley of the River Clyde, and in the former mining communities of Leadhills and Wanlockhead in the Lowther Hills.

The oldest rocks in the district occur within the Southern Uplands Terrane and consist of a thin succession of pillow lava and chert of Ordovician (Arenig–Caradoc age). These oceanic rocks are succeeded by a thick marine succession of wacke sandstone, siltstone and mudstone of Ordovician (Caradoc) and Silurian (Llandovery) age (Figure 1). These were deposited in the Iapetus Ocean as submarine fan turbidite sequences on a sea floor of pillow lava, chert and black mudstone. As the Iapetus Ocean closed by northward subduction the submarine fans were underthrust northwards beneath older packets of sediment. The sediments are now preserved as north-east–south-west orientated fault-bounded tracts of steeply dipping strata. Following closure of Iapetus, these sequences were buried, folded and uplifted. In the semi-arid conditions of late Silurian to Devonian times, the rocks underwent rapid erosion, giving rise to the red-bed successions of the Lower Old Red Sandstone (Lanark Group) of the Midland Valley Terrane (Figure 2). Volcanic activity was also widespread, producing lava flows and a variety of intrusions.

During the Mid-Devonian, deformation and uplift caused reactivation of several faults and resulted in a major unconformity between Devonian and Carboniferous strata. The Carboniferous rocks, including the Coal Measures, comprise cyclic sequences of marine limestone and mudstone, deltaic and fluviatile sandstone, seatearth and coals. In the Thornhill Basin, the lavas at the base of the Permian succession lie unconformably on Carboniferous strata. The lavas are succeeded by fluviatile sandstone and breccia, and red aeolian dune sandstone. The unconformity at the base of the Permian succession is an expression of the Hercynian orogeny, which resulted in folding and faulting in the district. Some of the faults were the focus for lead–zinc mineralisation, which occurred from Carboniferous to Permian times. Atlantic sea floor spreading in the Palaeogene produced several large north-west-trending dykes.

During the last major Pleistocene glaciation, ice flowed southwards along the Nith valley and northwards along the Clyde; it probably also radiated out from smaller centres in the Lowther Hills. Till deposited by ice occupies the main valleys and the lower ground in the north-west. As the ice melted, hummocky moraines and glacial outwash sand and gravel were deposited locally along the river valleys. Postglacial rivers reworked glacial sediments into alluvial terraces, and peat formed in poorly drained areas.

Figure 1 Lower Palaeozoic geological succession in the Southern Uplands Terrane.

System	Age	Group	Supergroup / Subgroup	Formation	Member	Lithology	Thickness
SILURIAN	Llandovery	GALA GROUP		GALA 7 UNIT		wacke sandstone	>2000 m
				Pibble Fault			
				GALA 6 UNIT		wacke sandstone	>2000 m
				Gillespie Burn Fault			
				GALA 4 UNIT		wacke sandstone	>2000 m
				Sandhead Fault			
				GALA 2 UNIT		wacke sandstone with some detrital pyroxene and/or hornblende	>2000 m
				Orlock Bridge Fault			
ORDOVICIAN	Caradoc-Ashgill			GLENLEE FORMATION		wacke sandstone with some detrital pyroxene and/or hornblende	>2000 m
				Glen Fumart Fault			
				SHINNEL FORMATION		wacke sandstone	>2000 m
				Fardingmullach Fault			
	Caradoc	SCAUR GROUP	LEADHILLS SUPERGROUP	PORTPATRICK FORMATION		volcaniclastic wacke sandstones	>2000 m
				Leadhills Fault			
				GALDENOCH FORMATION		volcaniclastic wacke sandstone	c.20 m
		BARRHILL GROUP		KIRKCOLM FORMATION	undivided	wacke sandstone	>3000 m
					SPOTHFORE MEMBER	breccia-conglomerate of sandstone, siltstone and chert clasts	up to 300 m
					STOODFOLD MEMBER	volcaniclastic sandstone, siltstone and conglomerate	c.10 m
					DUNTERCLEUCH CONGLOMERATE MEMBER	fossiliferous conglomerate including clasts of rotted calcareous sandstone	c.20 m
				Carcow Fault			
		TAPPINS GROUP		MARCHBURN FORMATION		wacke sandstone, siltstone, microconglomerate, chert and sedimentary breccia	>1000 m
				diachronous relationship			
SILURIAN	Llandovery	MOFFAT SHALE GROUP		UPPER BIRKHILL SHALE FORMATION		grey and black mudstone	>20 m
ORDOVICIAN-SILURIAN	Caradoc-Llandovery			LOWER BIRKHILL SHALE FORMATION		black mudstone	20 m
ORDOVICIAN	Caradoc			UPPER HARTFELL SHALE FORMATION		grey mudstone with black laminae	28 m
				LOWER HARTFELL SHALE FORMATION		black mudstone	22 m
				GLENKILN SHALE FORMATION		black cherty mudstone	20 m
	late Llanvirn (Llandeilian)-Caradoc	CRAWFORD GROUP		KIRKTON FORMATION		chert, siliceous mudstone, pillow lava	c.120 m
	late Arenig-Llanvirn			*stratigraphical gap?*			
	mid-Arenig			RAVEN GILL FORMATION	undivided	brown cherty mudstone, mudstone/chert mélange, pillow lava	55–70 m
					CASTLE HILL MEMBER	red cherty mudstone	7 m

Figure 2 Geological succession in the Midland Valley Terrane.

QUATERNARY	Holocene	Flandrian		Artifical (man-made) deposits Peat, alluvium including river terrace deposits and fans	
	Pleistocene	Devensian		Loch Lomond Stadial: hummocky glacial deposits	0–50 m
				Dimlington Stadial: till, glaciofluvial sand and gravel	

Intrusion of Palaeogene dyke suite
unconformity

UPPER PALAEOZOIC	LOWER PERMIAN		APPLEBY GROUP (Snar Valley Outlier)	GLENDOURAN BRECCIA FORMATION	reddish brown breccia conglomerate and sandstone	c.700 m
			APPLEBY GROUP (Thornhill Basin)	THORNHILL SANDSTONE FORMATION	bright red aeolian desert sandstone	c.200 m
				DURISDEER FORMATION	sandstone and sandy breccia	c.70 m
				CARRON BASALT FORMATION	olivine-basalt lavas	up to 20 m
	CARBONIFEROUS	West- phalian	COAL MEASURES GROUP	UPPER COAL MEASURES	mudstone, siltstone, sandstone, seatearth, and coal, reddened towards top	<100 m
				MIDDLE COAL MEASURES		300 m
				LOWER COAL MEASURES		200 m
		Namurian	CLACKMANNAN GROUP	PASSAGE FORMATION — undivided	sandstone, minor siltstone, mudstone and coal	up to 100 m
				PASSAGE FORMATION — TOWNBURN SANDSTONE MEMBER (Thornhill Basin)	sandstone and conglomerate	40 m
				ENTERKIN MUDSTONE FORMATION (Thornhill Basin)	reddened marine mudstone, siltstone, sandstone and seatearth	15 m
				UPPER LIMESTONE FORMATION	sandstone, siltstone, mudstone, marine limestone, coal and seatearth	up to 160 m
				LIMESTONE COAL FORMATION	sandstone, siltstone, mudstone, ironstone, coal and seatearth	140 m
				LOWER LIMESTONE FORMATION	mudstone, siltstone, sandstone, marine limestone	25 m
			STRATHCLYDE GROUP	LAWMUIR FORMATION	mudstone and sandstone	40 m
	Siluro-Devonian of the Midland Valley Terrane		LANARK GROUP	AUCHTITENCH SANDSTONE FORMATION	sandstone, conglomerate, siltstone, mudstone, local basalt lavas	1000 m
				DUNEATON VOLCANIC FORMATION	basalt, basaltic andesite and andesite lavas, volcaniclastic rocks	1200 m
				SWANSHAW SANDSTONE FORMATION	pinkish brown lithic sandstone	700 m

2 Geological description

Ordovician

With the exception of the north-west and south-east parts of the area, Ordovician rocks, part of the Southern Uplands Terrane, underlie most of the Leadhills district (Figure 3). The **Crawford Group (CRFD)**, which lies at the base of the succession, is composed of chert, mudstone and lava. It is divided into two formations between which no stratigraphical contact has been observed. The **Raven Gill Formation (RAVN)** consists of pillow lavas, brown and red mudstone and grey bedded chert. A conodont fauna gives a mid-Arenig age. The *Castle Hill Member* is a unit of red cherty mudstone interbedded with grey mudstone. Evidence of *mélange* lithologies and the presence of a stratigraphical gap within the Crawford Group suggest that the entire Raven Gill Formation may be an allochthonous olistostrome unit (Floyd, 2001). The succeeding or enclosing **Kirkton Formation (KRK)** consists of bedded chert and pillow lavas, with a conodont fauna of late Llanvirn to early Caradoc age. The Crawford Group metabasaltic rocks are part of a suite of Ordovician basaltic volcanic rocks which occur within the Northern Belt of the Southern Uplands. Geochemical studies indicate that these volcanic rocks were generated in tholeiitic and alkaline within-plate, mid-ocean ridge and volcanic arc settings (Phillips et al., 1995; Barnes et al., 1996). The Crawford Group is overlain by black and dark grey mudstone of the **Moffat Shale Group (MFS)** (Plate 1), dated as Caradoc to Llandovery on the basis of graptolite faunas. In the district, much of the mudstone is bleached and altered, probably as the result of hydrothermal fluid circulation during phases of mineralisation. Being largely within an imbricate zone, most Moffat Shale Group outcrops are highly tectonised.

The black mudstone is diachronously overlain by a thick sequence of Ordovician to Silurian marine turbidite wacke sandstone. This comprises, over most of the district, the Ordovician **Leadhills Supergroup** (and, in the south-east of the district, the Silurian Gala Group; see next section) (Figure 4). The **Tappins Group** is represented in the district by the **Marchburn Formation (MCHB)**, which occurs at the northern margin of the Southern Uplands Terrane and is bounded to the north by the Southern Upland Fault and to the south by the Carcow Fault. The formation youngs to the north-west and consists of turbiditic wacke sandstone, siltstone, chert and lenses of 'Haggis Rock' microconglomerate. Sedimentary structures indicate palaeocurrent directions orientated north-east–south-west axial to the basin margin with some flow from the south-east. South of the Carcow Fault and north of the Leadhills Fault, lies the thick **Kirkcolm Formation (KKF)** of the **Barrhill Group.** The Eller Fault separates the formation into two tracts of slightly different ages. The formation comprises well-bedded turbiditic sandstone which youngs predominantly north-westwards. Palaeocurrents indicate transport from the north-east or south-west. Intercalated within the Kirkcolm Formation are coarse-grained volcaniclastic sandstones and breccias of the *Stoodfold Member* (STOD) which appear to have been derived from the Bail Hill Volcanic Group in the New Cumnock district. The *Spothfore Member* (SPOT) is a breccia-conglomerate unit and the *Duntercleuch Conglomerate Member* comprises mass-flow deposits of fossiliferous conglomerate. The **Galdenoch Formation (GDF)** forms a 20 m-thick intercalation within the Kirkcolm Formation, and is composed of green-grey, pyroxene-bearing wacke sandstone.

The **Scaur Group** occupies the tract between the Leadhills and Fardingmullach faults and, in this district, is composed

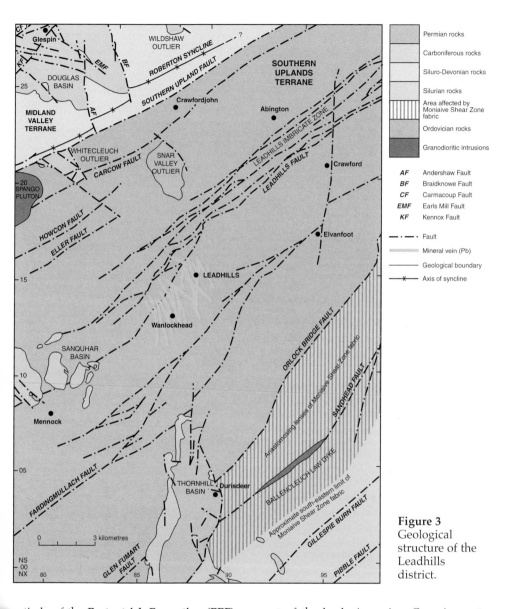

Figure 3
Geological structure of the Leadhills district.

entirely of the **Portpatrick Formation (PPF)**. The formation is dominated by volcaniclastic wacke sandstone, which contains abundant andesitic detritus with fresh detrital pyroxene and hornblende (Styles et al., 1989, 1995). In the Leadhills–Wanlockhead district, this sandstone forms the host country rock for most of the lead–zinc veins. Cropping out between the Fardingmullach and Glen Fumart/Orlock Bridge faults, the **Shinnel Formation (SHIN)** consists of wacke sandstone and thick units of laminated siltstone. To the south of the Glen Fumart Fault, the **Glenlee Formation (GNE)** is the youngest

Plate 1   View of Gripps Cleuch, Glengonnar Water, Leadhills. The broken line shows the contact between the bedded cherts of the Kirkton Formation (KRK), Crawford Group on the left and the Glenkiln Shale Formation (GBS), Moffat Shale Group on the right. *Inset* shows detail of contact (marked by stick), between chert on the left and black mudstone on the right. This is the proposed type section for the base of the Glenkiln Shales, and hence the Moffat Shale Group (Floyd, 2001). Scale: stick is 1.2 m long, but is lying on sloping ground.

Ordovician wacke sandstone unit in the Southern Uplands. It consists of interbedded quartzose and pyroxene-rich sandstone.

Silurian of the Southern Uplands Terrane

The south-east part of the district, to the south-east of the Orlock Bridge Fault, forms part of the Central Belt of the Southern Uplands. This area is occupied by wacke sandstone of the **Gala Group** of Llandovery age (Figure 5). The Gala Group comprises a series of north-east-trending, fault-bounded tectonostratigraphical tracts. Owing to the generally monotonous petrography of the group as a whole, individual tracts have not been erected as formal formations, but are instead defined by their position with respect to the tract-bounding faults. Biostratigraphical correlation between tracts is based largely on outcrops of the Moffat Shale Group outside the district. The **Gala 2** unit is the oldest and most north-westerly turbidite succession and is bounded by the Orlock Bridge Fault to the north-west and the Sandhead Fault and Ballencleuch Law

Figure 4 Ordovician Leadhills Supergroup.

Group	Formation	Lithology	Key localities	Graptolite biozones	Age
	GLENLEE	interbedded quartz-rich and pyroxene-rich wacke sandstone laminated siltstone	Corfardine, Scar Water [NX 810 962]	*anceps* Biozone (*pacificus* Subzone)	Ashgill
			Glen Fumart Fault		
	SHINNEL	thin- to medium-bedded turbiditic wacke sandstone and thick laminated siltstone	Motorway cutting at Fall Kneesand [979 161 to 983 160]; quarry at Glenochar [948 135]; Dinabid Linn [894 087]	By analogy with areas to SW, the formation rests on Moffat Shale Group with graptolites up to *linearis* Biozone	Caradoc–Ashgill
			Fardingmullach Fault		
SCAUR	PORTPATRICK	turbiditic wacke sandstone and laminated siltstone; rich in andesitic detritus	Scapcleuch Burn [915 170]; Long Cleuch [920 174]; Breckagh Burn [840 085]	Rests on Moffat Shale Group with graptolites of *gracilis* to *linearis* biozones	
			Leadhills Fault		
BARRHILL	GALDENOCH	massive pryoxene-bearing wacke sandstone	Quarry near Lettershaws [898 202]; Wandel Burn [969 262]		
	KIRKCOLM	quartz-rich wacke sandstone turbidites, silstones, graptolitic mudstones and chert beds. DUNTERCLEUCH CONGLOMERATE MEMBER: mass-flow deposits of conglomerate and coarse-grained sandstone. STOODFOLD MEMBER: volcaniclastic sandstones and breccias. SPOTHFORE MEMBER: breccia-conglomerate, sandstone and siltstone	Snar Water [864 205]; Bught Burn [849 206]; Duntercleuch Burn [8404 1605]; Brown's Cleuch [8102 1623]; Crawick Water [806 165]	*gracilis* Biozone	Caradoc
			Carcow Fault		
TAPPINS	MARCHBURN	turbiditic wacke sandstones and siltstones, lenses of micro-conglomeratic 'Haggis Rock', bedded chert	Stonehill Bank [830 209]; Duneaton Water [838 209]; Bught Burn [835 213]		

dyke to the south-east. It consists of wacke sandstone and siltstone, with a significant proportion of beds containing fresh detrital pyroxene and/or hornblende. Most of the Gala 2 tract is affected by the high-strain Moniaive Shear Zone (Barnes et al., 1995; Phillips et al., 1995). The **Gala 4** unit, lying south of Gala 2 between the Sandhead and Gillespie Burn faults, is compositionally similar to the non-pyroxenous parts of

Figure 5   Silurian Gala Group of the district. This group is made up of tectonostratigraphically defined units.

Group	Unit	Lithology	Key localities	Graptolite biozones	Age
GALA	GALA 7	quartzose wacke sandstone and siltstone	Capel Water [NX 970 992]	*turriculatus* Biozone (in Thornhill district)	Llandovery
	Pibble Fault				
	GALA 6	quartzose wacke sandstone and siltstone	Crow Craig [960 017]	Base of unit is *convolutus* Biozone (in Thornhill district)	
	Gillespie Burn Fault				
	GALA 4	wacke sandstone and siltstone, non-pyroxenous; partly affected by Moniaive Shear Zone	Carsehope Burn [954 049]	*triangulatus* Biozone (in Thornhill district)	
	Sandhead Fault				
	GALA 2	wacke sandstone and siltstone with some detrital pyroxene and/or hornblende; most of unit affected greatly by Moniaive Shear Zone	Kirk Burn [896 040]; Durisdeer Hill [915 051]; Comb Law [944 074]	At least *acuminatus* Biozone or younger (based on areas to NE and SW of district)	
	Orlock Bridge Fault				

Gala 2. **Gala 6** lies between the Gillespie Burn and Pibble faults and consists of quartzose wacke sandstone and siltstone. Only a very small area of **Gala 7** exists in the district, in the extreme south-east corner. Like Gala 6, the unit consists of quartzose wacke sandstone and siltstone.

Siluro–Devonian of the Midland Valley Terrane

The Siluro-Devonian rocks of the Midland Valley Terrane consist of about 3000 m of sedimentary and volcanic rocks, which together make up the **Lanark Group** (Figure 6). The group consists mainly of terrigenous clastic rocks with an intervening pile of calc-alkaline volcanic rocks. The **Swanshaw Sandstone Formation (SWAS)** consists of reddish brown lithic sandstone that was deposited in a relatively high-energy fluvial regime (Smith, 1993). The overlying **Duneaton Volcanic Formation (DNV)** forms part of a calc-alkaline volcanic suite that erupted on to continental crust in the southern part of the Midland Valley early in Devonian times. The formation is subaerial and composed of calc-alkaline basalt, basaltic andesite and andesite (Thirlwall, 1981b, 1982, 1983; Smith, 1993; Phillips, 1994, 2000). The volcanic rocks are locally intercalated with sandstone and conglomerate. The lavas may have been related to active continental margin volcanism above a former north-west-dipping subduction zone (Thirlwall 1981b, 1983) or, alternatively, they may have been derived by partial melting of a subduction-contaminated mantle source in an overall sinistral strike-slip regime (Smith, 1995; Phillips et al., 1998). The **Auchtitench Sandstone Formation (AUC)** includes several substantial conglomerate units, including the *Dungavel Hill Conglomerate Member* (DGHC). The conglomerates contain clasts of basaltic and andesitic lava in a sandstone matrix. Pebble imbrication in one conglomerate lens indicates that

Figure 6 Siluro-Devonian rocks in the Midland Valley Terrane.

Group	Formation	Lithology	Key localities	Age
LANARK	AUCHTITENCH SANDSTONE	fluviatile sandstone with subordinate conglomerate and rare lenses of basaltic and andesitic lavas. DUNGAVEL HILL CONGLOMERATE MEMBER: massive unsorted conglomerate with clasts of basalt and andesite	Duneaton Water [795 232 to 805 219]; Blackmire Burn [813 237]	Considered to be Early Devonian in age
	DUNEATON VOLCANIC	lavas of porphyritic basalt, basaltic andesite and andesite, with volcaniclastic sandstone, breccia and lapilli-tuff	Shieldholm area [808 213]; Greenfield Law area [877 251]; Sheriffcleuch Burn [9037 2112]	Radiometric ages of 411 to 407 Ma suggest an early Devonian age
	SWANSHAW SANDSTONE	red-brown fluviatile sandstone, cross-bedded in places, with subordinate conglomerate and siltstone	Not well exposed in the district but seen nearby in New Cumnock district [797 282]	

palaeocurrents flowed towards the north-north-west. Locally, the formation contains lenses of andesitic and basaltic lavas.

Carboniferous

Carboniferous rocks occur in relatively small basins or graben in the Midland Valley terrane, in the northern part of the district, and the southern parts of the Douglas Basin and the Wildshaw Outlier (Figure 7). Carboniferous rocks are restricted to the Whitecleuch Outlier and parts of the Sanquhar and Thornhill basins (Figures 8, 9) in the Southern Uplands terrane.

The **Strathclyde Group (SYG)** in the district is relatively thin and is represented by rocks of the **Lawmuir Formation (LWM)**. This formation lies at the base of the Carboniferous successions in the Douglas Basin and Wildshaw Outlier, where it rests unconformably on the Siluro-Devonian Lanark Group, and in the Whitecleuch Outlier where it is unconformable on Lower Palaeozoic rocks of the Marchburn Formation. The Lawmuir Formation comprises sandstone, mudstone and seatearths with thin marine limestones near the top. The **Clackmannan Group (CKN)** is undivided at the eastern margin of the Sanquhar Basin, where it is exposed as

remnant outliers, overlain unconformably by Lower Coal Measures. The rocks form a condensed, mostly marine, sequence which was partly eroded before the desposition of the Lower Coal Measures. At Howat's Burn [828 097], the sequence consists of a basal conglomerate that passes up into sandstone overlain by calcareous mudstone and limestone rich in shell fossils. In the Thornhill Basin (McMillan, 2002), the oldest Carboniferous succession is the Lower Carboniferous **Enterkin Mudstone Formation** which forms the base of the Clackmannan Group in this area. It consists of purplish mudstone and seatearth interbedded with red-brown siltstone and mudstone yielding fossil brachiopods, bivalves, nautiloids, orthocones and trilobite pygidia (Brand, 1990). The **Lower Limestone Formation (LLGS)** occurs in the Douglas Basin and the Wildshaw and Whitecleuch outliers. It contains a cyclic sequence of transgressive marine limestones and mudstones, which pass up into siltstones, sandstones and, in some cycles, seatearths and coals. The **Limestone Coal Formation (LSC)**, which occurs in the Douglas Basin and the Whitecleuch Outlier, is composed of upward-coarsening deltaic cycles of sandstone, seatearth, coal and mudstone, which are locally associated with thin marine bands.

Figure 7   Carboniferous rocks in the southern part of the Douglas Basin and the Wildshaw and Whitecleuch outliers.

Group	Formation	Lithology	Key localities	Age	
COAL MEASURES (Scotland)	UPPER COAL MEASURES	sandstone, siltstone, mudstone, with a few seatearths and coals, some reddening	Glespin [807 281]; scattered exposures in Thornhill Basin, e.g. [8821 0100], [8857 0161], [8857 0161]	Bolsovian	WESTPHALIAN
	MIDDLE COAL MEASURES	cyclical sequence of sandstone, siltstone, mudstone, with subordinate seatearths and coals	Auchenmulleran Burn [8707 0345]; River Nith [NX 8605 9981]; NX 8609 9971]	Duckmantian	
	LOWER COAL MEASURES	cyclical fluviodeltaic sandstone, siltstone, mudstone, seatearths and coals	tributary of Kennox Water [7976 2651]; Duneaton Water [8162 244]; Coal Burn [815 106]	Langsettian	
CLACKMANNAN	PASSAGE FORMATION	fluviatile sandstones, pebbly beds, thin red or green mudstones and a few thin coals and seatearths	Lees Hill [8018 2584]; Glentaggart Burn [8097 2453 to 8102 2491]; Kennox Water [7954 2633]		
	UPPER Limestone	deltaic cycles of sandstone, siltstone, mudstone, marine limestone, thin coal and seatearths. INDEX LIMESTONE at base	no good exposures in district; much inferred from boreholes	Pendleian	NAMURIAN
	LIMESTONE COAL	deltaic cycles of sandstone, seatearth, coal and mudstone, locally with thin marine bands	Mossy Burn [8091 2111]; Whitecleuch Quarry [8259 2070]; Coal Burn [8215 2152]		
	LOWER LIMESTONE	cyclic sequence of marine limestones (e.g. HURLET LIMESTONE, McDONALD LIMESTONE), deltaic mudstones, siltstones, sandstones, seatearths and coals	Glentaggart Burn [8127 2529] and 8092 2449]; Wildshaw Outlier [877 280]; Mossy Burn [8091 2108]	Visean	DINANTIAN
STRATHCLYDE	LAWMUIR	sandstone, siltstone, mudstone, seatearth and, near the top, thin marine limestones (e.g. DOUGLAS UNDER LIMESTONE)	Glentaggart Burn [8120 2520]; West Glespin [8156 2800]		

The Upper Limestone Formation (ULGS) is exposed only in the Douglas Basin, where it is represented by a cyclic deltaic sequence of sandstones, siltstones, mudstones, marine limestones, thin coals and seatearths. The **Passage Formation (PGP)** occurs in the Douglas and Thornhill basins. It consists of mainly buff to white sandstone with pebbly beds, red or green mudstone with thin coals and seatearths. The formation was deposited in an active fluvial environment. In the Thornhill Basin the formation is represented by the *Townburn Sandstone Member*, which consists of up to 15 m of grey, white and pink, coarse-grained sandstone lying unconformably on older Carboniferous or Lower

Figure 8   Permo-Carboniferous rocks in the Sanquhar Basin.

Group	Formation	Lithology	Key localities	Age
	CARRON BASALT	olivine-basalt lava	Small outcrop east of Sanquhar [797 091]	Lower Permian
		unconformity		
COAL MEASURES (Scotland)	MIDDLE COAL MEASURES	micaceous sandstone siltstone, mudstone, ironstone and thin coals	Ryehill Cleuch [7962 0895]	Duckmantian
	LOWER COAL MEASURES	sandstone, pebbly in places, siltstone, seatearth, mudstone, coal	Coal Burn [815 106]; Shiel Burn [820 113]	Langsettian
		unconformity		
CLACKMANNAN	Undivided	basal breccia, marine mudstone, siltstone, seatearth, coal	Howat's Burn [828 097] Red Burn [815 095] Bogs Burn [813 120]	Visean
		Unconformity on Ordovician rocks of the Leadhills Supergroup		

Palaeozoic strata. Up to 40 m of this member were proved in a borehole to the south of the district, and McMillan and Brand (1995) interpreted the despositional environment as deltaic, with low-sinuosity fluvial channels.

The Coal Measures include the coal-bearing, cyclic, fluviodeltaic beds that overlie the Passage Formation. They consist of sandstone, siltstone and mudstone with subsidiary seatearth and coals. The Lower and Middle Coal Measures were formerly termed the Productive Coal Measures and are overlain by the Upper (formerly Barren Red) Coal Measures. The **Lower Coal Measures (LCMS)** occur in the Whitecleuch Outlier and the Sanquhar and Thornhill basins. The rocks are middle and upper Langsettian in age. The Vanderbeckei Marine Band marks the base of the **Middle Coal Measures (MCMS)** which occurs in the Douglas, Sanquhar and Thornhill basins. The strata are Duckmantian in age and the top of the formation is taken at the base of the Aegiranum Marine Band. The **Upper Coal Measures (UCMS)** are limited to the Douglas and Thornhill basins. The coals are generally of poor quality as a result of local reddening of strata. The reddening of Carboniferous strata in the Thornhill Basin was ascribed by Simpson and Richey (1936) to weathering processes in the Permian.

Lower Permian

The major outcrop of Permian strata in the district is in the Thornhill Basin where Carboniferous and Permian rocks occupy a fault-bounded basin (Figure 3). The Lower Permian **Carron Basalt Formation (CnB)** rests unconformably on Coal Measures and is composed of up to 20 m of thin alkali basalt lava flows interbedded with stream-flood breccias and sandstones. The breccias contain basalt and wacke sandstone clasts. The lavas are interpreted as the products of small-scale rifting in late Carboniferous to early Permian times. This formation also crops out towards the eastern margin of the Sanquhar Basin. In the Thornhill Basin, the lavas are unconformably overlain by the **Durisdeer Formation (DU)**, which consists of breccia dominated by basalt clasts with interbedded sandstone and siltstone. The formation is interpreted as a sequence of sheet flood desert floor sediments (Brookfield, 1980). The overlying **Thornhill Sandstone Formation (THHS)** is composed of red aeolian dune sandstone with well-developed cross-stratification.

Permian continental clastic rocks also crop out in an outlier at Snar valley where they form the **Glendouran Breccia Formation**. This outlier may be the remnant of the clastic infill of a half-graben developed during the Permian. This formation

Figure 9   Permo-Carboniferous rocks in the Thornhill Basin.

Group	Formation	Lithology	Key localities	Age	
APPLEBY	THORNHILL SANDSTONE	red, dune-bedded sandstone	Carron Water [855 027]	LOWER PERMIAN	
	DURISDEER	breccia, fluviatile sandstone and siltstone	Type section in Hapland Burn [888 023 to 889 025]		
	unconformity				
	CARRON BASALT	olivine-basalt lava, stream-flood breccias and sandstones	Type section in Carron Water [885 017 to 888 022]; Jenny Hair's Bridge [886 024]		
	unconformity				
COAL MEASURES (SCOTLAND)	UPPER COAL MEASURES	sandstone, siltstone, mudstone	Commonly not present; minor exposures near Carron Water [8821 0100]; Kiln Cleuch [8857 0161]	Bolsovian	WEST-PHALIAN
	MIDDLE COAL MEASURES	reddened mudstone, siltstone, seatearth and sandstone	River Nith [NX 8605 9981; NX 8609 9971]; Auchenmulleran Burn [8707 0345]	Duckmantian	
	LOWER COAL MEASURES	sandstone, purple and red siltstone and mud-stone, mottled seatearth	River Nith [NX 8607 9975]	Langsettian	
CLACKMANNAN	PASSAGE	represented by pebbly fluvial sandstones of the Townburn Sandstone Member	River Nith [NX 859 999]; Morton Mains [NX 887 996]	NAMURIAN–WEST-PHALIAN	
	unconformity				
	ENTERKIN BURN	purple-red sandstone, seatearth, fossiliferous siltstone and mudstone	Type section in Enterkin Burn [8696 0520 to 8724 0540]; Glenvalentine Burn [8793 0774]	DINANTIAN–NAMURIAN	

rests unconformably on Ordovician rocks and comprises reddish brown breccia-conglomerate. Clasts are derived mainly from the underlying Lower Palaeozoic strata. Variable palaeocurrents suggest that local drainage may have infilled the developing half-graben.

Intrusive igneous rocks

Ordovician metagabbroic rocks are the oldest intrusive rocks in the district and are closely associated with, and probably intruded into, the metabasaltic volcanic rocks of the Crawford Group. The metagabbros comprise plagioclase laths with intergranular clinopyroxe and possible olivine pseudomorphs. Locally, these intrusive rocks are intensely deformed, forming a suite of cataclasites and mylonites.

Many dykes in the district are part of a regional swarm of **Caledonian minor intrusions** (Barnes et al., 1986; Rock et al., 1986a, b; Rock et al., 1988). These include andesite/microdiorite, dacite/microgranodiorite and rhyolite/microgranite, and were originally known as the 'porphyrite–porphyry' series (Greig, 1971). Calc-alkaline lamprophyres (including both spessartites and kersantites) are also present (Rock, 1984). Although no age dates are available from the Leadhills district, lamprophyre dykes in the Kirkcudbright district have been dated at 395 to 418 Ma.

The early Devonian granodiorite and diorite intrusions in the district include part

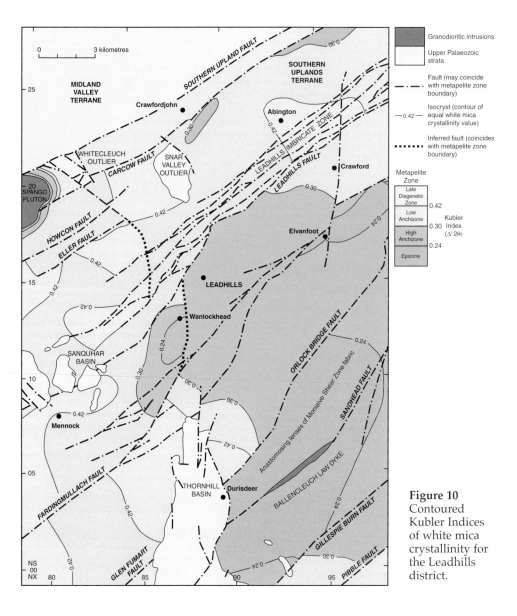

Figure 10
Contoured
Kubler Indices
of white mica
crystallinity for
the Leadhills
district.

of the Spango granodiorite and the smaller, elongate Ballencleuch Law granodiorite. The Spango intrusion is composed of hornblende-biotite-granodiorite and hornblende-bearing biotite-granodiorite. The intrusion has produced a thermal aureole of hornfels within surrounding sedimentary country rocks. The Ballencleuch Law granodiorite intrusion is poorly exposed and is represented typically by a north-east-trending boulder field.

Permo-Carboniferous minor intrusions in the district comprise large sills of basic alkaline rocks intruded into Carboniferous sedimentary rocks of the Midland Valley. These are mainly composed of alkali olivine-dolerites (microgabbro). The Midland Valley also contains a suite of tholeiitic quartz-dolerite dykes and sills, which trend east–west and cut Devonian and Coal Measures rocks. Emplacement of both the alkali dolerite and tholeiitic quartz-dolerite intrusions into coal seams resulted in the development of 'white trap' in which the rock is altered to a white or brownish kaolinite/carbonate-rich assemblage.

An unusual porphyritic alkali micro-gabbro, the Crawfordjohn 'Essexite', is exposed in Craighead Quarry [918 237]. The intrusion takes the form of a north-west-trending dyke, 15 to 25 m wide. It contains large augite phenocrysts and was formerly used to make curling stones.

Palaeogene minor intrusions form part of a high level, north-west-trending regional dyke swarm of tholeiitic micro-gabbro and basalt. These dykes occur throughout western and central Scotland and were intruded 55 to 61 Ma ago, during the opening of the Atlantic Ocean. A large example can be seen in a quarry south-west of Stoneyburn [9594 1933], where a 16 m-thick dyke of tholeiitic microgabbro, with chilled margins, is exposed.

Regional structure and metamorphism

The Southern Upland Fault, which sepa-rates the Midland Valley and Southern Uplands terranes, has had a long history of movement from the Early Palaeozoic through Devonian and into post-Westphalian times. Through its history it has acted with strike-slip, normal and reverse senses of movement (Floyd, 1994). Ultimately, it acted as the south-eastern bounding fault of the Midland Valley graben. The two terranes had different tec-tonic and sedimentary histories in the Early Palaeozoic but, since the closure of the Iapetus Ocean and certainly from the Late Devonian onwards, they have broadly shared the same deformation events.

Caledonian deformation was related to closure of the Iapetus Ocean from the late Ordovician to mid-Silurian. Although there is debate about the overall setting of the Southern Uplands Terrane, with models ranging from a forearc trench (McKerrow et al., 1977; Hepworth et al., 1982) to a backarc basin (Stone et al., 1987), it is apparent that the terrane represents an imbricated thrust stack of marine sandstones and siltstones. These Lower Palaeozoic sedimentary rocks were deposited on large submarine fans on the southern margin of the Laurentian conti-nent. As the Iapetus Ocean closed, probably by northward subduction of oceanic crust under Laurentia, packets of submarine fan sediments were underthrust northwards. This process was repeated many times, with each newly accreted slice thrust northwards under its predecessor. The entire stack also gradually rotated northwards towards the vertical, and was overturned in places. The original *décollement* surfaces between the accreted packets now form tract-bounding faults; in the district these include the Carcow, Leadhills, Fardingmullach, Glen Fumart, Orlock Bridge, Sandhead, Gillespie Burn and Pibble faults. As a result of these processes, the rocks generally young north-wards within any one tract, but successive tracts towards the south include sequentially younger sandstones. Folds observed within the tracts tend to have two main orientations: those with subhorizontal axes and steep north-east-trending axial planes (F_1–F_2 folds), and those with near-vertical axes and north-east-trending axial planes (F_3 folds). The former may relate to orthogonal closure of Iapetus, whereas the F_3 folds may have formed in a later, sinistral, strike-slip regime when the two continents had become closer.

The **Acadian deformation** of mid-Devonian times resulted in the unconformity between the Lanark and Strathclyde groups, although the unconformity may also partly relate to thermal doming associated with extrusion of the early Visean Clyde Plateau lavas in the western part of the Midland Valley. During the mid-Devonian, the Southern Upland Fault had a northerly downthrow component, and drag on the fault produced the steep dips observed in folded Lanark Group rocks directly to the north.

Hercynian deformation relates to syn- and postdepositional faulting during the Carboniferous on reactivated Caledonian (north-east trend) faults and also new, north- to north-north-west-trending faults. The former led to the development, between the Carmacoup and Kennox faults, of a narrow graben that contains the main syncline of the Douglas Coalfield. The southern part of the coalfield, lying south of the Kennox Fault, consists of a series of north-east-tilted half-graben developed against the north-north-west trending fault set. The north- and north-north-west-trending fault sets were also active through into the Permian and form the bounding faults to the east-tilted half graben of the Thornhill Basin and Snar Valley Outlier as well as the full graben of the Whitecleuch Outlier. Faults on these north- and north-north-west trends also acted as conduits for the base-metal mineralisation of the Leadhills–Wanlockhead mining district.

Regional metamorphism has produced particular clay mineral assemblages in the Lower Palaeozoic mudstones of the district. These can be analysed and used to interpret the metamorphic history of the region. Measurements of clay mineral reaction progress, particularly white mica (illite) crystallinity, have allowed Merriman and Roberts (2001) to produce a contoured metamorphic map which delineates zones of diagenesis and low-grade metamorphism in the imbricated Ordovician and Silurian strata (Figure 10). This map shows a pattern that is characterised by isocrysts (contours of equal crystallinity) that are subparallel with the regional strike, and abrupt changes in grade across the strike-parallel, tract-bounding faults. This is a common pattern across the Southern Uplands and indicates a close relationship between the imbrication of the succession and the regional metamorphism. This regional pattern, which was probably produced as a result of accretionary burial, is locally overprinted by the contact aureole of the Spango intrusion and localised shear zone metamorphism associated with the Moniaive Shear Zone. A system of north-north-west-trending faults through the centre of the district, associated with the development of Permo-Carboniferous basins, has also modified the regional pattern.

Concealed geology

Information on the concealed geology of the district is provided by boreholes and mine plans (mainly from the coalfields and the Leadhills–Wanlockhead lead-mining area) and also from the interpretation of geophysical data. A brief description of the main geophysical features is given here.

The north-western part of the **gravity anomaly** map (Figure 11) is dominated by a north-eastward-trending Bouguer gravity anomaly high. The gravity gradient on the north side of this feature coincides with the Southern Upland Fault and can be explained, at least in part, by the presence of relatively low density Upper Palaeozoic rocks on the north side of the fault. The gravity gradient on the south side of the high is probably due to concealed structures. Seismic profiles show that basement to the south of the Southern Upland Fault has a high density, similar to that of the Midland Valley (Kimbell et al., in press). This may indicate the presence of an extension of the concealed basement of the Midland Valley (Hall et al., 1983; Davidson et al., 1984). Northward anomaly trends are dominant in the south-eastern part of the gravity map. A gravity low with this trend occurs over the Thornhill Basin and results from the low density of the basin-fill compared with the underlying Lower Palaeozoic rocks. The north–south contours on the eastern side of the district may be related to the effect of a concealed feature to the east of the district. One possibility is the postulated concealed 'Tweeddale Granite' batholith (Lagios and Hipkin, 1971; Stone et al., 1997).

Magnetic anomaly data show a contrast between the magnetic signatures of the northern and southern parts of the district (Figure 12). The northern part is dominated by short wavelength magnetic anomalies,

Figure 11   Bouguer gravity anomaly map of the Leadhills district.
Reduction density = 2.72 Mg/m³. Contour interval = 0.5 mGal; colour
interval = 1 mGal. Crosses indicate gravity stations.

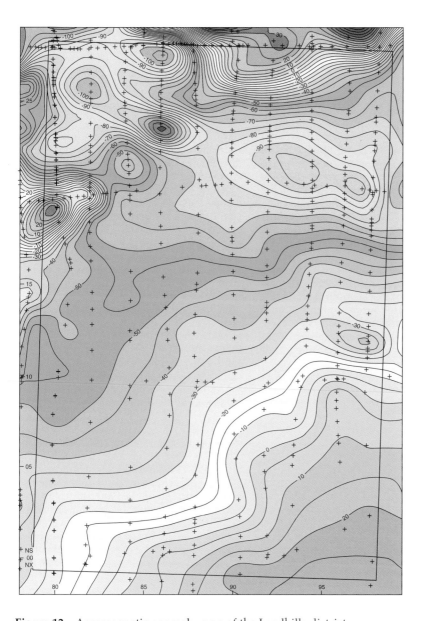

Figure 12   Aeromagnetic anomaly map of the Leadhills district.

Total field anomalies relative to a local variant of IGRF90. Contour
interval = 5 nT; colour interval = 10 nT. Crosses indicate locations of
digitised points.

both positive and negative, associated with near-surface sources. These sources probably include concealed Devonian volcanic rocks of the Midland Valley (Smith, 1999a) and magnetic sedimentary units in the Marchburn Formation (Floyd and Trench, 1989; Smith, 1999a), both giving rise to positive anomalies. A positive magnetic anomaly south of the Southern Upland Fault is associated with the Spango intrusion to the west, and a belt of south-east-trending Palaeogene dykes gives rise to a series of negative magnetic anomalies. The main effect farther south is the rise in magnetic field towards the south-eastern corner of the district. This feature is the flank of a major long-wavelength magnetic high (the 'Galloway High' of Powell, 1970). Kimbell and Stone (1995) suggested that this anomaly is due to a zone of relatively magnetic mid-crustal rocks bounded to the south by the Iapetus suture and to the north by a basement structure which was reactivated to produce the Moniaive Shear Zone.

Quaternary

Most Quaternary deposits and landforms in the district probably date from the last major glaciation, which occurred during the Dimlington Stadial (28 000 to 13 000 years BP) of the Late Devensian Stage. During this cold period, valley glaciers in the Highlands, Southern Uplands and Lake District grew and coalesced into an ice-sheet which, at its maximum, covered the whole of Scotland and northern England. To the north of the district, in the Midland Valley, south-flowing ice from the Highlands converged with north-flowing ice from the Southern Uplands and the resulting sheet moved to flow generally eastwards. The Lowther Hills acted as a barrier to the eastward movement of the main ice stream so that some of the ice was diverted south-eastwards down the Nith valley and the rest flowed north-eastwards along the southern margin of the Midland Valley. The Lowther Hills also acted as a local centre of ice dispersion and fed a northwards-flowing ice stream along the valley of the Clyde to the east

of the Lowther Hills which merged with the Midland Valley ice stream.

Till was deposited beneath the ice, forming a layer of sandy diamicton in the valleys and on the lower ground of the north-west. The character of the till varies according to the local bedrock. For example, the Lower Palaeozoic rocks give rise to grey, sandy, stony deposits whereas the Carboniferous rocks produce a tenacious blue-grey clayey till. Areas of Devonian and Permian strata are charaterised by red sandy till. Two distinct types of till can be distinguished: a lower, compact subglacial till laid down beneath an active glacier, and an overlying loose, coarser grained supraglacial ablation till deposited during glacial retreat. Most pebbles and boulders within the tills are of local origin.

Glaciofluvial sand and gravel is mainly crudely stratified glacial outwash material that forms mounds, terraces and ridges along some of the major valleys, particularly the Nith and Carron waters. Late-glacial marginal meltwater channels cut in both rock and till are well developed in the Nith valley below Enterkinfoot and in the Carron valley around Durisdeer [894 038].

Hummocky glacial deposits (moraine) are preserved in the upper reaches of the Clyde and its tributary the Elvan and are thought to have been deposited by local glaciers growing in the Lowther Hills during the Loch Lomond Readvance (around 10 500 years BP). The deposits consist of ridges and mounds of poorly sorted, earthy, unstratified glacial debris along the flanks of valleys.

Flandrian (postglacial and present-day) deposits accumulated in the warmer, wetter climate that began about 10 000 years BP at the beginning of the Holocene. By this time most of Scotland was probably ice-free and drainage patterns similar to those of the present-day were beginning to become established. **Alluvium** occurs along the floodplains of major rivers such as the Clyde, Duneaton and Nith and also along many secondary watercourses such as the Wanlock, Snar, Glengonnar (Plate 2), Elvan, Carron, Glespin and Daer. The material is mostly sand and silt with lenses of gravel

Plate 2   View of the lower reaches of the Glengonnar Water, Abington showing the once extensive drape of till which formerly filled the main valley, but is now eroded by the meandering river to form a terrace feature. Castle Hill, left background, lies in the centre of the Leadhills Imbricate Zone and has extensive exposures of lava, chert and sheared black mudstone on its western slopes (P507306).

and layers of peat. In the larger valleys, such as the Clyde, the alluvium is present as one or more terraces. Isolated patches of silty alluvium (e.g. around Kirkhope [963 054]) may represent lacustrine deposits laid down in late glacial or postglacial lakes. Small alluvial fans are common at the junctions between steeply graded side streams and main valleys. Good examples occur in the upper reaches of the Dalveen Pass [900 076].

Head (soliflucted hillwash) occurs as a veneer over many of the hills in the district. It is a loose earthy layer of unsorted, frost-shattered rock with a silty or clayey matrix that formed in periglacial conditions during the Late Devensian. It is typically less than 1 m thick, but is locally thicker where solifluction lobes have crept downslope. Head is stabilised by vegetation in most areas but erodes rapidly on slopes where plant cover is breached. Such erosion is common on the hillsides around Wanlockhead and Leadhills where miners used surface water to flush away surface deposits such as head in order to expose mineral veins.

Peat occurs as blanket peat on hill tops and also as flat deposits covering some of the large alluvial spreads in the main valleys. Much of the hilltop peat is drying out and being eroded away by grazing animals, wind and rain. Peat is mapped only where it exceeds 1 m in thickness.

Landslips have occurred, typically, where river meanders have undercut terraces of unconsolidated drift deposits, and are generally of small scale.

3 Applied geology

Parts of the Leadhills district have had a long history of mineral extraction, principally coal around Glespin and lead–zinc in the Leadhills–Wanlockhead area. Mining for both has now ceased.

Mineral resources

The only significant **coal** resources in the district are the seams in the Lower and Middle Coal Measures of the southern part of the Douglas Coalfield. There are twelve persistent coals in the Lower Coal Measures, most of which have been worked in the Glespin area, principally by mining but some by opencast. There are also three workable coals in the lower part of the Middle Coal Measures; these were formerly exploited in the Kennox–Glespin area.

The principal areas of **sand and gravel** in the district occur along the Nith and Carron Water valleys. They are not known to have been worked on any scale. Other limited areas of sand and gravel occur around Glespin [80 27] and along Mill Burn [90 27]. Some of the former may overlie potential open-cast coal sites.

A range of local rock types has been used as **building stone**. The numerous dry stone dykes of the district were generally built using stone from adjacent outcrops, including Carboniferous sandstone, microgabbro (dolerite), Lower Palaeozoic greywacke sandstone or any suitable loose blocks from the drift. The attractive **Crawfordjohn 'Essexite'** from Craighead quarry was formerly worked on a limited scale for the manufacture of curling stones. The quarry could potentially be reopened for monumental or other decorative stone production. **Limestone**, from the Hurlet Limestone, was formerly quarried in the Glespin area and in the Wildshaw Outlier. Much of the latter locality has now been grassed over or covered by material derived from the M74 construction. The Glenochar 'slate' quarry south of Elvanfoot [948 135] is in laminated siltstone of the Shinnel Formation and was formerly worked for an inferior slate. Some of the spoil heaps have been removed for bulk fill in the past and could potentially supply more. As a roofing slate the rock is relatively worthless. There is a substantial **mine spoil** heap of broken rock at the former Glencrieff Shaft at Wanlockhead, which has been used over the years for bulk fill material. However, the historical importance of this heap in terms of the lead mining industry means that its future use for bulk fill is questionable.

The Leadhills–Wanlockhead mining district contains over 70 **lead–zinc–silver veins** (Figure 13). The veins are typically banded fissure-fillings, formed during two periods of mineralisation. They cut across Palaeozoic rocks south of and within the Leadhills Imbricate Zone and comprise two distinct sets: one set striking north-north-west and the other striking north–south. About 0.4 million tonnes of metallic lead, 10 000 tonnes of zinc and 25 tonnes of silver were obtained from this area between 1700 and 1958. This area has also produced several tonnes of **gold** from alluvial sources since the mid-16th century. Lead veins have also been recorded to the north of the mining district, at Glendouran [888 205] and Gilkerscleuch [90 21 and 91 22], and alluvial gold has been found in association with Permian rocks in the Thornhill Basin (Leake and Cameron, 1996). The most comprehensive accounts of the Leadhills– Wanlockhead lead–zinc veins, their contents and historic workings have been produced by Porteous (1876), Wilson (1921), Mackay (1959), Temple (1956) and Gillanders (1981). Location details of most of the old workings are provided by Landless (1993). The exposures described in Borthwick's (1993) excursion guide provide a good overview of the geology and mineralisation of the mining district. The mineral veins

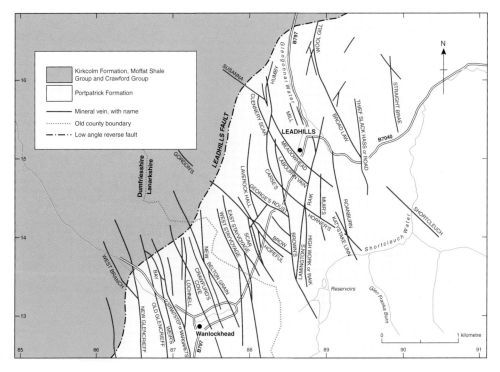

Figure 13   Map of the principal named veins in the Leadhills–Wanlockhead Mining District.

in the Leadhills–Wanlockhead mining district are also of great interest for the wide range of minerals, 57 in all, that they contain. The mining district is the type locality for lead-hillite [$Pb_4SO_4(CO_3)_2(OH)_2$], lanarkite [Pb_2SO_5] and caledonite [$Cu_2Pb_5(SO_4)_3 CO_3(OH)_6$] (Plate 3). Recent exploration in the mining district has focused on the search for the bedrock source of the alluvial gold. Recent research has established that there are likely to have been several distinct sources for the alluvial gold (Leake et al., 1997).

Groundwater

Groundwater flow in Lower Palaeozoic rocks is largely controlled by fractures and fault planes, since the lithified nature of the turbiditic sandstones means that intergranular permeability is almost zero. The resultant localised channelling of groundwater produces springs at surface and the formation of recharge zones to groundwater where river valleys intersect faults at surface. Areas underlain by Lower Palaeozoic rocks are generally of high topographic relief and limited population. Hence, groundwater exploitation is generally limited to a small number of boreholes, wells and springs. Most springs are fed from the weathered zone immediately below rockhead, which owes its enhanced hydraulic conductivity to dilated fissures that allow water to flow at a higher rate than at depth. In the past, mining operations in the Wanlockhead–Leadhills area were hampered by inflowing water, which entered the mines from surface watercourses. A range of drainage adits, river diversions and culverts were constructed to alleviate the problem. However, the surface water was also utilised where possible, to power pumps to drain the mines. Ditches were constructed to intercept run-off and store it in reservoirs. A variety of waterwheels and engines were powered by this means. Groundwater from Lower Palaeozoic strata is of variable quality, some contains high concentrations of iron and manganese that is unacceptable for human consumption.

Lower Devonian sandstones in the north of the district are poorly known hydrologically. In other areas these strata have a low to moderate hydraulic conductivity and are characterised by fissure flow. These characteristics are therefore assumed to be similar for the Leadhills district. Carboniferous strata in the north of the district contain moderately permeable aquifers. Groundwater was a significant problem during coal mining in the Douglas Coalfield, and mine pumping had to be carried out. Permian aeolian sandstone near Durisdeer [895 035] is the most permeable rock unit in the district. Intergranular permeability, measured less than 500 m south of the district, indicates that, unlike the Devonian and Carboniferous aquifers, intergranular groundwater flow in the Permian rocks is more significant than fissure flow (Ball, 1999). Water quality from the Permian aquifer is excellent. This, together with the fact that the Permian sandstone occupies low-lying ground where recharge is high, indicates that resource potential for this unit is significant.

Fluvioglacial and alluvial gravels in the main valleys generally have a high permeability and shallow water table. Their role in absorbing surface runoff and floodwater in the rivers may be significant. These aquifers have not been exploited for groundwater but the potential exists.

Geology and planning

Geological assessments play an important role in planning for land-use development. They can provide details of the resources available and the main types of geological hazards likely to be encountered. There is

Plate 3   Crystals of blue caledonite $[Pb_5Cu_2(CO_3)SO_4)_3(OH)_6]$ on grey / black ?anglesite, Leadhills. Specimen Number MI30346 (MNS 4495).

an increasing demand to manage developments which affect the visual impact of the landscape, and even to initiate landscape improvements. There is also a growing awareness of the value of geological heritage in the rock outcrops and the landscape, not only to the geological community for scientific study, but also as part of the tourist attraction. These interests need to be balanced against the continuing need for essential mineral resources and the provision of land for housing, commercial, industrial, waste disposal and other developments.

No **minerals** are currently being exploited in the district, although there may be potential for opencast coal working and hardrock quarrying. Rock, till and sand and gravel generally provide sound **foundation conditions** below the top weathered zone. Poor foundation conditions can be caused by

Plate 4   Remains of the former wood-lined flues from the smelter at Wanlockhead. The bare soil on the surrounding hillside testifies to the extent of heavy-metal contamination in the vicinity (Z00402).

superficial deposits at surface, such as peat, clay and silt, alluvial deposits, and landfill. Foundation conditions may also be affected by variably compressible buried superficial deposits like peat and soft clay and by liquefaction of silt and fine sand. All these deposits require careful site investigation.

Abandoned mineworkings for coal, limestone, ironstone and lead–zinc present a hazard where collapse of workings may propagate to the surface. The main threat in the coal mining areas is pillar or roof failure in stoop and room workings, particularly where they are within 30 to 40 m of the surface. Potential foundation problems are possible within, but not limited to, areas underlain by Lower and Middle Coal Measures and Limestone Coal Formation. Shafts and adits present localised hazards.

The Leadhills–Wanlockhead former mining district presents its own particular problems from old mineworkings, particularly as the sites of many of the shafts have been lost or obscured by the spoil from later workings. Apart from shafts, the main danger is from the collapse of stoped areas in the veins. There is also a potential problem in this area of the presence of lead-contaminated water, soil and dust from centuries of mining and smelting (Plate 4).

The stability of drift deposits on steep slopes may be affected by loading, natural undermining by meandering rivers and/or excavation. These processes make them susceptible to minor landslip and debris flow. The stability of bedrock in cliffs and steep-sided excavations may depend on its resistance to weathering and the presence and orientation of joints, faults and inclined bedding planes. Movement on such planar features may give rise to rockfalls as well as landslips.

Discharges of ferruginous water from deep mines can be a problem after mine closure when dewatering operations cease. In the district, ferruginous drainage has caused **pollution of surface watercourses** in the Glentaggart area (Wood, 1995) where the local colliery formerly pumped large amounts of groundwater from a depth of 87 m in the Lower and Middle Coal Measures. Discharges such as these can have serious deleterious effects on the ecology of local watercourses.

Former quarries and disused or active sand and gravel workings are potential sites for **landfill**. Landraise, in which natural hollows and valleys are infilled, would also be possible in the district.

Man-made ground falls into three main categories. *Made ground*, such as tips of waste rock, is common near former mines. Removal of these tips for bulk fill and to recover remaining mineral content, especially coal, may be beneficial within the district. Motorway road embankments are also classified as made ground. *Worked ground* is mapped where excavations have been made in the original ground surface; this category includes cuttings for motorways and railways. Former open-cast coal sites, which have been excavated and backfilled, are designated as *worked and infilled ground*.

Geological heritage

Geotourism is a significant source of income in the Leadhills–Wanlockhead mining district and is centred on the Museum of Lead Mining at Wanlockhead.

Information sources

Further geological information held by the British Geological Survey relevant to the Leadhills district is listed below. Searches of indexes to some of the collections can be made on the Geoscience Data Index in BGS libraries and on the BGS web site at http://www.bgs.ac.uk.

Maps

1:625 000
Solid geology map UK North Sheet, 2001; Quaternary geology, 1977

1:250 000
Sheet 55N 04W Borders, Solid geology, 1986

1:63 360
Sheet 16 Moffat, Solid and Drift 1924 (out of print)
Sheet 10 Dumfries, Solid and Drift 1885 (out of print)
Sheet 23 Hamilton, Solid 1929 (out of print); Drift 1929 (out of print)

1:50 000
Sheet 9W New Galloway, Solid 1978; Drift 1979
Sheet 9E Thornhill, Solid 1996; Drift 1980
Sheet 10W Lochmaben, Drift 1983
Sheet 15W New Cumnock, Solid 1999; Solid and Drift 1999
Sheet 23W Hamilton, Solid 1995; Drift 1993
Sheet 24W Biggar, Solid 1980; Drift 1981

The Leadhills district is the eastern half of the area which was formerly covered by Sheet 15 (Sanquhar) of the Geological Map of Scotland.

1:10 000 and 1:10 560
Details of the original geological surveys are listed on editions of the 1:63 360 and 1:50 000 Series geological sheets. Copies of the fair-drawn maps of these earlier surveys may be consulted at the BGS Library, Edinburgh.

The maps at 1:10 560 (six-inch to one mile) or 1:10 000 scale covering the 1:50 000 Series Sheet 15E are listed below, together with the sheet name, surveyors' initials and the dates of the survey. The surveyors were H F Barron, I B Cameron, J D Floyd, A A McMillan, E A Pickett and R A Smith. The maps are not published but are available for consultation in the Library, BGS Edinburgh, and also at BGS Keyworth and the BGS London Information Office, in the Natural History Museum, South Kensington, London. Photocopies may be purchased from the BGS Sales Desk.

Map No	Map name	Surveyor	Date
NS70NE*	Sanquhar	RAS	1996
NS70SE*	Glenwhargen	HFB	1995
NS71NE*	Kirklea	RAS	1996
NS71SE*	Crawick	RAS	1996
NS72NE*	Glenbuck	IBC	1990
NS72SE*	Auchendaff	RAS	1993
NS80NW	Mennock	EAP	1997
NS80NE	Enterkin Path	EAP	1997
NS80SW	Fardingmullach	EAP	1997
NS80SE	Durisdeer	EAP	1997
NS81NW	Spango	RAS	1998
NS81NE	Leadhills	JDF	1998
NS81SW	Glendyne	EAP	1999
NS81SE	Wanlockhead	EAP	1999
NS82NW*	Glespin	IBC	1996
NS82NE*	Wildshaw	IBC	1996
NS82SW	Mosscastle	RAS	1997
NS82SE	Crawfordjohn	RAS	1998
NS90NW	Dalveen Pass	JDF	1997
NS90NE*	Daer	JDF	1997
NS90SW	Ballencleuch Law	JDF	1997
NS90SE*	Daer Water	JDF	1997
NS91NW	Elvan Water	JDF	1999
NS91NE*	Elvanfoot	HFB	1998
NS91SW	Green Lowther	EAP	1999
NS91SE*	Tormont Hill	EAP	1999
NS92NW*	Roberton	HFB	1999
NS92NE*	Wandel	HFB	1997
NS92SW	Abington	JDF	1999
NS92SE*	Crawford	HFB	2000
NX79NE*	Arkland Rig	JDF	1993
NX89NW*	Druidhall	AAM	1994
NX89NE*	Thornhill	AAM	1991
NX99NW*	Gatelawbridge	AAM	1991
NX99NE*	Queensberry	AAM	1991

* part sheet

Digital geological map data

In addition to the printed publications noted on the previous page, many BGS maps are available in digital form, which allows the geological information to be used in GIS applications. These data must be licensed for use. Details are available from the Intellectual Property Rights Manager at BGS Keyworth. The main datasets are:

DiGMapGB-625 (1:625 000 scale)
DiGMapGB-250 (1:250 000 scale)
DiGMapGB-50 (1:50 000 scale)
DiGMapGB-10 (1:10 000 scale)

The current availability of these can be checked on the BGS web site at:

http://www.bgs.ac.uk/products/digi-talmaps/digmapgb.html

Geophysical maps

1:1 5000 000
Colour shaded relief gravity anomaly map of Britain, Ireland and adjacent areas, 1997. Colour shaded relief magnetic anomaly map of Britain, Ireland and adjacent areas, 1998.

1:250 000
Sheet 55N 04W   Borders, Aeromagnetic anomaly, 1980; Bouguer gravity anomaly 1991.

1:50 000
Geophysical Information Maps are plot-on-demand maps which summarise graphically the publicly available geophysical information held for the district in the BGS databases. Features include regional gravity and aeromagnetic data, location of geophysical surveys, location of public domain seismic surveys and the location of deep boreholes.

Geochemical atlases

1:250 000
Southern Scotland and part of Northern England, 1993. Point-source geochemical data are processed to generate a smooth continuous surface presented as an atlas of small-scale colour-classified digital maps.

Hydrogeological maps

1:625 000
Sheet 18 (Scotland) 1988.
Groundwater vulnerability (Scotland) 1995.

Books

British Regional Geology
Midland Valley of Scotland. Third edition, 1985
South of Scotland. Third edition, 1971

Memoirs and Sheet Explanations
Geology of the New Cumnock district (Sheet 15W), 1999
Geology of the New Galloway and Thornhill district (Sheets 9W and 9E), 2002
Explanation of Sheet 15, 1871
Geology of the Hamilton district (Sheet 23W), 1998

Sheet Descriptions
Geology of the Leadhills district (Sheet 15E), 2002
Geology of the New Cumnock district (Sheet 15W), 1999

Documentary collections

Borehole record collection
BGS holds collections of records of boreholes which can be consulted at BGS, Edinburgh, where copies of most records may be purchased. For the Leadhills district, the collection consists of the sites and logs of about 1172 boreholes. Index information for these boreholes has been digitised. Many of the older records are drillers' logs.

Site exploration reports
This collection consists of site exploration reports carried out to investigate foundation conditions prior to construction. There is a digital index and the reports are held on microfiche. At the time of writing there are about 18 reports for the Leadhills district.

Mine plans
BGS maintains a collection of plans of underground mines for minerals other than coal and oilshale. This collection comprises 147 plans for lead-zinc workings.

Hydrogeological data
Records of water boreholes are held at BGS, Edinburgh.

Geochemical data
Records of stream-sediment and other analyses are held at BGS, Keyworth.

Gravity and magnetic data

Records are held at BGS, Keyworth.

Seismic data

Records of British earthquakes are held at BGS, Edinburgh.

Material collections

Photographs

About 50 photographs illustrating aspects of the geology of the Leadhills district are deposited for reference at BGS libraries in Edinburgh and Keyworth, and in the BGS Information Office, London. The photographs were taken between 1937 and 1991 and depict details of the various rocks and sediments exposed either naturally or in excavations, and also some general views. A list of titles can be supplied on request. Colour or black and white prints and transparencies can be supplied at a fixed tariff from the Photographic Department, BGS, Edinburgh.

Petrological collections

The petrological collections for the Leadhills district consist of over 200 hand specimens and thin sections. Although many samples and thin sections are of the igneous rocks in the district, the wacke sandstones of the Southern Uplands Terrane are also well represented. Information on databases of rock samples, thin sections and geochemical analyses can be obtained from the Mineralogy and Petrology Section, BGS, Edinburgh.

Borehole core collection

Rock samples have been collected from core taken from selected boreholes and are held in the petrological collection. Details can be obtained from the Mineralogy and Petrology Section, BGS, Edinburgh.

Palaeontological collections

The collections of biostratigraphical specimens are taken from surface and temporary exposures, and from boreholes throughout the Leadhills district. The collections are working collections and are used for reference. They are not at present on a computer database. Macrofossils are held in the collection at BGS, Edinburgh from where details of fossil samples from the Leadhills district can be obtained from the Curator, Palaeontology Section, BGS, Edinburgh.

Other relevant data

Coal abandonment plans are held by the Coal Authority, Mining Records Department, Bretby Business Park, Ashby Road, Burton-on-Trent, Staffordshire, DE15 0QD.

Sites of Special Scientific Interest are the responsibility of Scottish Natural Heritage, Battleby Redgorton, Perth, PH1 3EW.

References

Most of the references listed below are held in the libraries of the British Geological Survey at Murchison House, Edinburgh and at Keyworth, Nottingham. Copies of the references can be purchased subject to the current copyright legislation.

ARMSTRONG, H A, OWEN, A W, SCRUTTON, C T, CLARKSON, A N K, and TAYLOR, C M. 1996. Evolution of the Northern Belt, Southern Uplands controversy. *Journal of the Geological Society of London*, Vol. 153, 197–205.

BALL, D F. 1999. An overview of groundwater in Scotland. *British Geological Survey Technical Report*, WD/99/44.

BARNES, R P, ROCK, N M S, and GASKARTH, J W. 1986. Late Caledonian dyke-swarms in Southern Scotland: new field, petrological and geochemical data for the Wigtown Peninsula, Galloway. *Geological Journal*, Vol. 21, 101–125.

BARNES, R P, PHILLIPS, E R, and MERRIMAN, R J. 1996. Allochthonous Ordovician basaltic rocks of possible island arc affinity in the Southern Uplands, SW Scotland. New Perspectives in the Appalachian-Caledonian Orogen. HIBBARD, J P, VAN STAAL, C R, and CAWOOD, P A (editors). *Geological Association of Canada*, Special Paper. No. 41.

BARNES, R P, PHILLIPS, E R, and BOLAND, M P. 1995. The Orlock Bridge Fault in the Southern Uplands of southwestern Scotland: a terrane boundary? *Geological Magazine*, Vol. 132, 523–529.

BLUCK, B J. 1978. Sedimentation in a late orogenic basin: the Old Red Sandstone of the Midland Valley of Scotland. 249–278 in Crustal evolution in northwestern Britain and adjacent regions. BOWES, D R, and LEAKE, B E (editors). *Special Issue of the Geological Journal*, No. 10.

BORTHWICK, G W. 1993. Leadhills and Wanlockhead. 192–202 in *Scottish Border geology: an excursion guide*. MCADAM, A D, CLARKSON, E N K and STONE P (editors). (Edinburgh: Scottish Academic Press).

BRAND, P J. 1990. Pre-Westphalian biostratigraphy of the Thornhill Basin (3). Report on collections made in the field from locations on sheets NX89NE, 89SE, 99SW, NS80SE, 80NE, forming part of the area of the Thornhill Basin, Sheets 9, 15, Scotland. *British Geological Survey Technical Report*, WH/90/272R.

BROOKFIELD, M E. 1980. Permian intermontane basin sedimentation in southern Scotland. *Sedimentary Geology*, Vol. 27, 167–194.

DAVIDSON, K A S, SOLA, M, POWELL, D W, and HALL, J. 1984. Geophysical model for the Midland Valley of Scotland. *Transactions of the Royal Society of Edinburgh: Earth Sciences*, Vol. 75, 175–181.

DAVIES, A. 1970. Carboniferous rocks of the Sanquhar Outlier. *Bulletin of the Geological Survey of Great Britain*, No. 31, 1–49.

FLOYD, J D. 1982. Stratigraphy of a flysch succession: the Ordovician of west Nithsdale, SW Scotland. *Transactions of the Royal Society of Edinburgh: Earth Sciences*, Vol. 73, 1–9.

FLOYD, J D. 1994. The derivation and definition of the 'Southern Upland Fault': a review of the Midland Valley–Southern Uplands terrane boundary. *Scottish Journal of Geology*, Vol. 30, 51–62.

FLOYD, J D. 1996. Lithostratigraphy of the Ordovician rocks in the Southern Uplands: Crawford Group, Moffat Shale Group, Leadhills Supergroup. *Transactions of the Royal Society of Edinburgh: Earth Sciences*, Vol. 86, 153–165.

FLOYD, J D. 2001. The Southern Uplands Terrane: a stratigraphical review. *Transactions of the Royal Society of Edinburgh: Earth Sciences*, Vol. 91, 349–362.

FLOYD, J D, and TRENCH, A. 1989. Magnetic susceptibility contrasts in Ordovician greywackes of the Southern Uplands of Scotland. *Journal of the Geological Society of London*, Vol. 146, 77–83.

GEIKIE, A, PEACH, B N, JACK, R L, SKAE, H, and HORNE, J. 1871. Explanation of Sheet 15. Dumfriesshire (north-west part); Lanarkshire (south part); Ayrshire (south-east part). *Memoir of the Geological Survey, Scotland*.

GILLANDERS, R J. 1981. Famous mineral localities: the Leadhills–Wanlockhead district. *The Mineralogical Record*, July–August, 235–250.

HALL, J, POWELL, D W, WARNER, M R, EL-ISA, ADESANYA, Z H M, and BLUCK, B J. 1983. Seismological evidence for shallow crystalline basement in the Southern Uplands of Scotland. *Nature*, Vol. 305, 418–420.

HEPWORTH, B C, OLIVER, G J H, and MCMURTRY, M J. 1982. Sedimentology, volcanism, structure and metamorphism of the northern margin of a Lower Palaeozoic accretionary complex; Bail Hill–Abingdon area of the Southern Uplands of Scotland. 521–534 *in* Trench–forearc geology. LEGGETT. J K (editor). *Geological Society of London Special Publication*, No. 10.

KIMBELL, G S, CARRUTHERS, R M, WALKER, A S D, and WILLIAMSON, J P. in press. Southern Scotland–Northern England. British Geological Survey, Geophysics CD Series.

LAGIOS, E, and HIPKIN, R G. 1979. The Tweeddale Granite — a newly discovered batholith in the Southern Uplands. *Nature*, Vol. 280, 672–5.

LANDLESS, J G. 1993. *A gazetteer to the metal mines of Scotland*. Third edition. Occasional Paper No.1. (Wanlockhead: The Wanlockhead Museum Trust).

LEAKE, R C, CHAPMAN, R J, BLAND, D J, CONDLIFFE, E, and STYLES, M T. 1997. Microchemical characterization of gold from Scotland. *Transactions of the Institution of Mining and Metallurgy*. (Section B: Applied earth sciences), Vol. 106, B85–98.

LUMSDEN, G I. 1964. The Limestone Coal Group of the Douglas Coalfield, Lanarkshire. *Bulletin of the Geological Survey of Great Britain*, No. 21, 37–71.

LUMSDEN, G I. 1965. The base of the Coal Measures in the Douglas Coalfield, Lanarkshire. *Bulletin of the Geological Survey of Great Britain*, No. 22, 80–91.

LUMSDEN, G I. 1967a. The Carboniferous Limestone Series of Douglas, Lanarkshire. *Bulletin of the Geological Survey of Great Britain*, No. 26, 1–22.

LUMSDEN, G I. 1967b. The Upper Limestone Group and Passage Group of Douglas, Lanarkshire. *Bulletin of the Geological Survey of Great Britain*, No. 27, 17–48.

LUMSDEN, G I, and CALVER, M A. 1958. The stratigraphy and palaeontology of the Coal Measures of the Douglas Coalfield, Lanarkshire. *Bulletin of the Geological Survey of Great Britain*, No. 15, 32–70.

MACKAY, R A. 1959. The Leadhills–Wanlockhead Mining District. 49–62 in *Symposium on the future of non-ferrous mining in Great Britain and Ireland*. (London: Institution of Mining and Metallurgy.)

MCGIVEN, A. 1967. Sedimentation and provenance of post-Valentian conglomerates up to and including the basal conglomerate of the Lower Old Red Sandstone in the southern part of the Midland Valley of Scotland. University of Glasgow, PhD thesis (unpublished).

MCKERROW, W S, LEGGETT, J K, and EALES, M H. 1977. Imbricate thrust model of the Southern Uplands of Scotland. *Nature*, Vol. 267, 237–239.

MCMILLAN, A A. 2002. Geology of the New Galloway and Thornhill district. *Memoir of the British Geological Survey*, Sheet 9 (Scotland).

MCMILLAN, A A, and BRAND, P J. 1995. Depositional setting of Permian and Upper Carboniferous strata of the Thornhill Basin, Dumfriesshire. *Scottish Journal of Geology*, Vol. 31, 43–52.

MCMURTRY, M J. 1980. Ordovician rocks of the Bail Hill area, Sanquhar, South Scotland: Volcanism and sedimentation in the Iapetus Ocean. PhD Thesis, University of St Andrews.

PEACH, B N, and HORNE, J. 1899. The Silurian rocks of Britain. Vol. 1: Scotland. *Memoir of the Geological Survey of the United Kingdom.*

PHILLIPS, E R. 1994. Whole-rock geochemistry of the calc-alkaline Old Red Sandstone lavas, Sheet 15W (New Cumnock), Lanarkshire, Scotland. *British Geological Survey Technical Report*, WG/94/1.

PHILLIPS, E R. 2000. Petrology of the igneous rocks exposed in the Leadhills district (Sheet 15E) of the Southern Uplands, Scotland. *British Geological Survey Technical Report*, WA/00/67.

PHILLIPS, E R, BARNES, R P, MERRIMAN, R J, and FLOYD, J D. 1995. The tectonic significance of Ordovician basaltic rocks in the Southern Uplands, SW Scotland. *Geological Magazine*, Vol. 132. 549–556.

PHILLIPS, E R, SMITH, R A, and CARROLL, S. 1998. Strike-slip, terrane accretion and the pre-Carboniferous evolution of the Midland Valley of Scotland. *Transactions of the Royal Society Edinburgh: Earth Sciences*, Vol. 89, 209–224.

PORTEOUS, J M. 1876. *God's treasure house in Scotland*. (London: Simpkin, Marshall and Co.)

POWELL, D W. 1970. Magnetised rocks within the Lewisian of Western Scotland and under the Southern Uplands. *Scottish Journal of Geology*, Vol. 6, 353–369.

ROCK, N M S, COOPER, C, and GASKARTH, J W. 1986a. Late Caledonian subvolcanic vents and associated dykes in the Kirkcudbright area, Galloway, south-west Scotland. *Proceedings of the Yorkshire Geological Society*, Vol. 46, 29–37.

ROCK, N M S, GASKARTH, J W, HENNEY, P J, and SHAND, P. 1988. Late Caledonian dyke-swarms of northern Britain: some preliminary petrogenetic and tectonic implications of their province-wide distribution and chemical variation. *Canadian Mineralogist*, Vol. 26, 3–22.

SIMPSON, J B, and RICHEY, J E. 1936. The geology of the Sanquhar Coalfield and adjacent basin of Thornhill. *Memoir of the Geological Survey, Scotland.*

SMITH, R A. 1993. Explanation for 1:10 000 Sheet NS72SE (Auchendaff). *British Geological Survey Technical Report*, WA/93/34.

SMITH, R A. 1995. The Siluro-Devonian evolution of the southern Midland Valley of Scotland. *Geological Magazine*, Vol. 132, 503–13.

SMITH, R A. 1999a. Geology of the New Cumnock district. *Sheet Description of the British Geological Survey*, 1:50 000 Series Sheet 15W New Cumnock (Scotland).

SMITH, R A. 1999b. Geology of the New Cumnock district — a brief explanation of the geological map. *Sheet Explanation of the British Geological Survey*, 1:50 000 Sheet 15W New Cumnock (Scotland).

STONE, P, FLOYD, J D, BARNES, R P, and LINTERN, B C. 1987. A sequential back-arc and foreland basin thrust duplex model for the Southern Uplands of Scotland. *Journal of the Geological Society of London*, Vol. 144, 753–764.

STONE, P, KIMBELL, G S, and HENNEY, P. 1997. Basement controls on the location of strike-slip shear in the Southern Uplands of Scotland. *Journal of the Geological Society of London*, Vol. 154, 141–144.

STYLES, M T, PEREZ-ALVAREZ, M, and FLOYD, J D. 1995. Pyroxenous greywackes in the southern Uplands and their petrotectonic implications. *Geological Magazine*, Vol. 132, 539–547.

STYLES, M T, STONE, P, and FLOYD, J D. 1989. Arc detritus in the Southern Uplands: mineralogical characterisation of a 'missing' terrane. *Journal of the Geological Society of London*, Vol. 146, 397–400.

SYBA, E. 1989. The sedimentation and provenance of the Lower Old Red Sandstone Greywacke Conglomerate, Southern Midland Valley, Scotland. Unpublished PhD thesis, University of Glasgow.

TEMPLE, A K. 1956. The Leadhills–Wanlockhead lead and zinc deposits. *Transactions of the Royal Society of Edinburgh*, Vol. 63, 85–114

THIRLWALL, M F. 1981. Implications for Caledonian plate tectonic models of chemical data from volcanic rocks of the British Old Red Sandstone. *Journal of the Geological Society of London*, Vol. 138, 123–138.

THIRLWALL, M F. 1982. Systematic variation in chemistry and Nd-Sr isotopes across a Caledonian calc-alkaline volcanic arc; implications for source materials. *Earth and Planetary Science Letters*, Vol. 58, 27–50.

THIRLWALL, M F. 1983. Isotope geochemistry and origin of calc-alkaline lavas from a Caledonian continental margin volcanic arc. *Journal of Volcanology and Geothermal Research*, Vol. 18, 589–631.

WILSON, G V. 1921. Special reports on the mineral resources of Great Britain; Vol. 17, The lead, zinc, copper and nickel ores of Scotland. *Memoir of the Geological Survey.*

WOOD, S C. 1995. An assessment of the temporal persistence of minewater pollution in the Midland Valley of Scotland. Unpublished MSc thesis. University of Newcastle, Water Resources Engineering, Department of Civil Engineering